Alim Nisa
Sajawal Ashraf
Hamood Rehman

Plant Stress Physiology 2

Alim Nisa
Sajawal Ashraf
Hamood Rehman

Plant Stress Physiology 2

Noor Publishing

Imprint

Any brand names and product names mentioned in this book are subject to trademark, brand or patent protection and are trademarks or registered trademarks of their respective holders. The use of brand names, product names, common names, trade names, product descriptions etc. even without a particular marking in this work is in no way to be construed to mean that such names may be regarded as unrestricted in respect of trademark and brand protection legislation and could thus be used by anyone.

Cover image: www.ingimage.com

Publisher:
Noor Publishing
is a trademark of
Dodo Books Indian Ocean Ltd., member of the OmniScriptum S.R.L Publishing group
str. A.Russo 15, of. 61, Chisinau-2068, Republic of Moldova Europe
Printed at: see last page
ISBN: 978-620-3-85955-3

ARTICLE

ON

Plant Stress Physiology
(Abiotic & Biotic Stresses in Plants)

WRITTEN

BY

MRS. ALIM-UN-NISA

M SAJAWAL ASHRAF

HAMOOD-UR-REHMAN

FOOD AND BIOTECHNOLOGY RESEARCH CENTER
PCSIR LABORATORIES COMPLEX, FEROZEPUR ROAD, LAHORE-54600

1

Table of Contents

Table of Figures

Introduction

What is Stress?

"Stress is the body's method of reacting to a condition such as a threat, challenge or physical and psychological barrier".

Stress, either physiological or biological, is an organism's response to a stressor such as an environmental condition. Stimuli that alter an organism's environment are responded to by multiple systems in the body. Environmental stresses play crucial roles in the productivity, survival and reproductive biology of plants as well as crops. Plants are subjected to many forms of environmental stress, which can be included into two broad areas: abiotic (physical environment) and biotic (e.g. pathogen, herbivore). However, plants evolve different mechanisms of tolerance to cope with the stress effects. These mechanisms comprise physiological, biochemical, molecular and genetic changes. This chapter represents a general overview of the major mechanisms developed by plants to tolerate environmental stresses, both abiotic (drought, high temperature, chilling and freezing, UV-B radiation, salinity and heavy metals) and biotic (herbivory, pathogen and parasite and allelopathy).

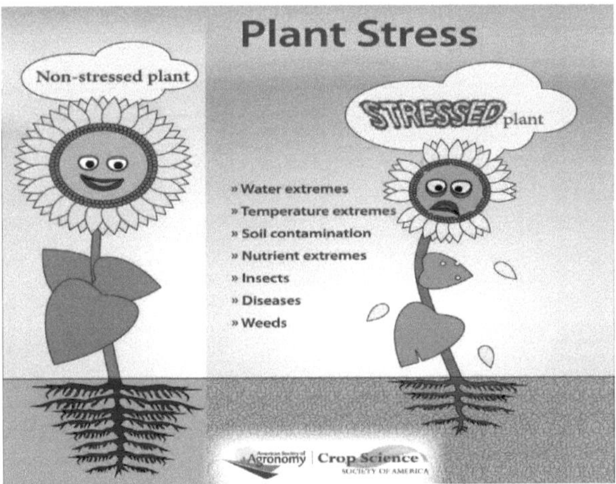

Figure 1: plant in stressed state

<u>**Plant Stress**</u>

"Plant stress is a state where the plant is growing in non-ideal growth conditions that increase the demands made upon it".

OR

"Stress in plants refers to external conditions that adversely affect growth, development or productivity of plants".

<u>**Effects of plant stresses:**</u>

* ❖ Stresses trigger a wide range of plant responses like altered gene expression, cellular metabolism, changes in growth rates, crop yields, etc.
* ❖ A plant stress usually reflects some sudden changes in environmental condition. The effects of stress can lead to deficiencies in growth, crop yields, permanent damage or death if the stress exceeds the plant tolerance limits.
* ❖ However, in stress tolerant plant species, exposure to a particular stress leads to **acclimation** to that specific stress in a time-dependent manner.

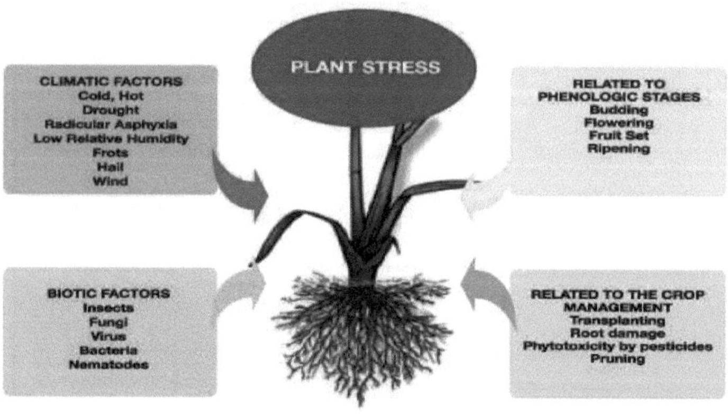

Figure 2: Plant stresses involving different climatic factors

❖ Some stresses to the plants injured them as such that plants exhibit several metabolic dysfunctions.

❖ The plants can be recovered from injuries if the stress is mild or of short term as the effect is temporary while as severe stresses lead to death of crop plants by preventing flowering, seed formation and induce senescence.

❖ Such plants will be considered to be stress susceptible. However, several plants like desert plants (Ephemerals) can escape the stress altogether

The basic concepts of plant stress, acclimation, and adaptation

Energy is an absolute requirement for the maintenance of structural organization over the lifetime of the organism. The components of stress acknowledgment and cell flagging. Briefly:plasma layers are the sensors of ecological changes; phytohormones andsecond couriers are the transducers of data from layers tometabolism; carbon balance is the expert integrator of plant reaction; betwixtand between, a few qualities are communicated all the more firmly, while others arerepressed. Responsive oxygen species assume key parts in all over guideline ofmetabolism and construction.

The maintenance of such complex order over time requires a constant through put of energy. The results in a constant flow of energy through all biological organisms, which provides the dynamic driving force for the performance of important maintenance processes such as cellular biosynthesis and transport to maintain its characteristic structure and organization as well as the capacity to replicate and grow. The maintenance of a steady-state results in a meta-stable condition called **homeostasis**.

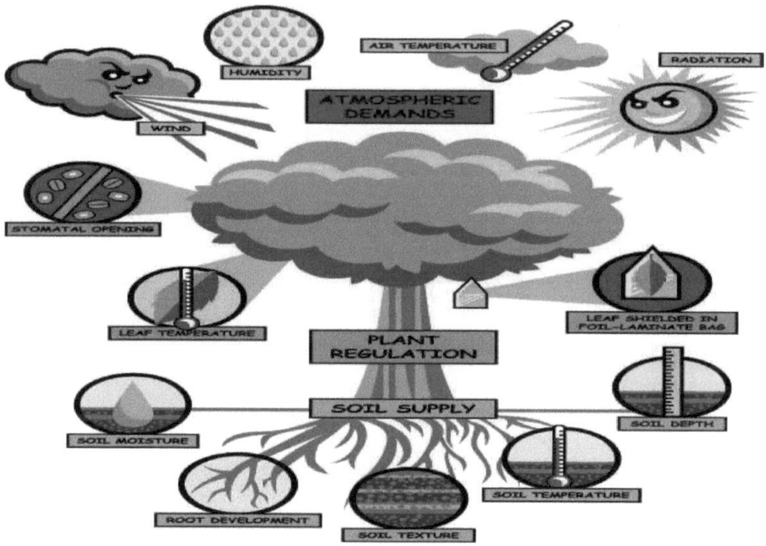

Figure 3: Plant physiology modulation of homeostasis

Environmental modulation of homeostasis defined as biological stress

Any change in the surrounding environment may disrupt homeostasis. Environmental modulation of homeostasis may be defined as **biological stress**. Thus, it follows that **plant stress** implies some adverse effect on the physiology of a plant induced upon a sudden transition from some optimal environmental condition where homeostasis is maintained to some suboptimal condition which disrupts this initial homeostatic state. Thus, plant stress is a relative term since the experimental design to assess the impact of a stress always involves the measurement of a physiological phenomenon in a plant species under a suboptimal, stress condition compared to the measurement of the same physiological phenomenon in the same plant species under optimal conditions.

Types of stresses:

Plants respond to stress in several different ways. Plant stress can be divided into two primary categories.

1. Biotic stresses

2. Abiotic stresses

Abiotic stresses in Plants

Introduction:

Plants are multicellular autotrophic living organisms in the kingdom Plantae that use photosynthesis to make their own food. They produce most of the world's oxygen, and are important in the food chain, as many organisms eat *plants* or eat organisms which eat *plant*. However, plants plays an important role in our ecosystem, but they are subjected to wide range of environmental stresses which reduces and limits the productivity of agricultural crops. Environmental stresses are divided into two groups that is one is biotic stresses and other is abiotic stresses. Here, we discuss the abiotic stresses of plants as follows:

"Abiotic stress may be defined as the negative impact of non-living factor on the living organisms in a specific environment".

The non-living variable must influence the environment beyond its normal range of variation to adversely affect the population performance or individual physiology of the organism in a significant way. Abiotic stress factors or stressors are naturally occurring and comes in many form. The most common stressors are easiest for people to identify, but there are many other, less recognizable abiotic stress factors which affect environment constantly.

Abiotic stresses in plants:

Abiotic stresses in plants is basically a physical (temperature, light etc.) or chemical insult that the environment may impose on the plants are termed as abiotic stresses in plants. Abiotic stresses in plants includes following factors such as

a. Heavy Metals/Metalloids stress
b. Drought (water stress)
c. Excessive watering (water logging)
d. Extreme temperatures (cold, frost and heat)
e. Salinity and mineral toxicity
f. Lesser known stressors are generally occurs on a small scale. They include pH, high radiation, compaction, contamination etc.

These abiotic stresses negatively impact growth, development, yield and seed quality of crop and other plants. In future it is predicted that fresh water scarcity will increase and ultimately intensity of abiotic stresses will increase.

It is a foremost factor that causes the loss of major crop plants worldwide. This situation is going to be more rigorous due to increasing desertification of world's terrestrial area, increasing salinization of soil and water, shortage of water resources and environmental pollution. Hence, there is an urgency to develop crop varieties that are resilient to abiotic stresses to ensure food security and safety in coming years.

Abiotic stresses are interconnected with each other and may occur in form of osmotic stress, malfunction of ion distribution and plant cell homeostasis. The growth rate and productivity is affected by a response caused by group of genes by changing their expression patterns.

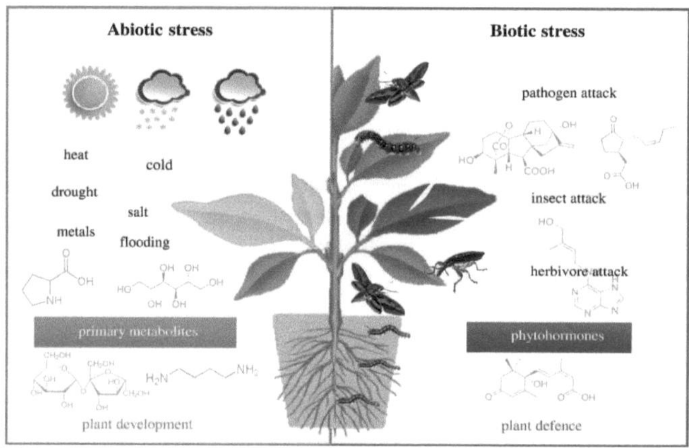

Figure 4: Abiotic stresses in plants

<u>Some main Abiotic Stresses in Plants</u>

> A plant's first line of defence against abiotic stress is in its roots. If the soil holding the plant is healthy and biologically diverse, the plant will have a higher chance of surviving stressful conditions.

➢ The plant responses to stress are dependent on the tissue or organ affected by the stress. For example, transcriptional responses to stress are tissue or cell specific in roots and are quite different depending on the stress involved.

➢ One of the primary responses to abiotic stress such as high salinity is the disruption of the Na+/K+ ratio in the cytoplasm of the plant cell. High concentrations of Na+, for example, can decrease the capacity for the plant to take up water and also alter enzyme and transporter functions. Evolved adaptations to efficiently restore cellular ion homeostasis has led to a wide variety of stress tolerant plants.

➢ Facilitation, or the positive interactions between different species of plants, is an intricate web of association in a natural environment. It is how plants work together. In areas of high stress, the level of facilitation is especially high as well. This could possibly be because the plants need a stronger network to survive in a harsher environment, so their interactions between species, such as cross-pollination or mutualistic actions, become more common to cope with the severity of their habitat.

➢ Plants also adapt very differently from one another, even from a plant living in the same area. When a group of different plant species was prompted by a variety of different stress signals, such as drought or cold, each plant responded uniquely. Hardly any of the responses were similar, even though the plants had become accustomed to exactly the same home environment.

Heavy Metals/Metalloids Stress

Plants to a great extent rely upon soil answer for obtain supplements for their development and formative cycle. The new expansion in pollution of arable terrains with substantial metals is one of the main sources of misfortune in crop efficiency (Proshad et al., 2018). Broad openness to weighty metal pollution undermines the supportability of natural and horticultural frameworks. Yields are regularly exposed to metal poisonousness because of ill-advised water system techniques and the expansion of extreme amounts of compound composts, and other engineered supplements (Pasala, 2017). Modern and sewage garbage removal, metropolitan overflow, consuming of strong and fluid fills, homegrown waste disposal in streams and trenches, and numerous other pathways cause substantial metal defilement. A few (conceivably poisonous) weighty metals, like Cu, Zn, Ni, Co, Se, and Fe, are likewise fundamental components needed for the ideal presentation of plants and become harmful when gathered in abundance in soil arrangement (Khan et al., 2018; Narendrula-Kotha et al., 2019). Then again, superfluous components, like arsenate (As), cesium (Cs), lead (Pb), and cadmium (Cd), can hamper crop efficiency when collected in the dirt even in follow sums (Khalid et al., 2018). Soil tainting with weighty metals causes gathering of these harmful metals in plant parts, bringing about diminished yield usefulness and expanded danger to creature and human wellbeing (Couto et al., 2018).

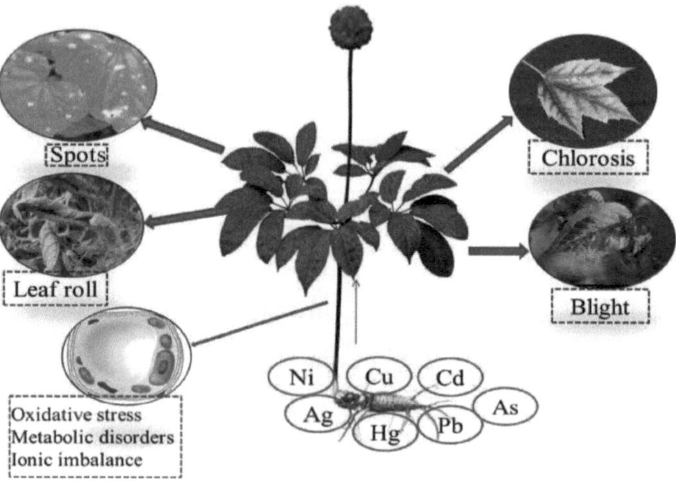

Figure 5: Heavy Metals/Metalloids Stress in Plants.

Arsenic (As) is a cancer-causing component present worldwide that destructive to each type of life (Ghosh and Mukhopadhyay, 2019). TU arranges redox homeostasis to give resilience against weighty metal pressure. TU-interceded redox homeostasis set off the downregulation of Lsi2 (carriers for as movement) which decreased as levels by 56%. Simultaneously, up-guideline of Sultr1; 1 and 1; 2 (sulfate carriers) improved sulfur digestion to enhance the unfavorable impacts of as stress (Srivastava et al., 2014). Through these systems, TU confines weighty metal movement to shoots and sequesters them in roots. One more conceivable system of diminished stacking of metal particles in xylem from roots to shoots with TU supplementation is chelation of metal particles with thiols (TU contains thiol-bunch) in the cytosol. Presence of thiols works with proficient use of SH-containing compounds as biomarkers of metal resistance (Borisova et al., 2016; Meng et al., 2019). Thiols likewise license vacuolar compartmentation (Guo et al., 2012), which may be a fundamental instrument through which TU diminishes metal-poisonousness, however this has not yet been tentatively demonstrated.

<u>Drought</u>

Nowadays climate has changed all around the globe by continuously increase in temperature and atmospheric CO2 levels. The distribution of rainfall is uneven due to the change in climate which acts as an important stress as drought.

The soil water available to plants is steadily **decreased** due severe drought conditions and cause death of plants prematurely. After drought is imposed on crop plants growth arrest is the first response subjected on the plants. Plants reduce their growth of shoots under drought conditions and reduce their metabolic demands. After that protective compounds are synthesized by plants under drought by mobilizing metabolites required for their osmotic adjustment.

Figure 6: Drought water stress in plants

One interesting thing that has been found in plants that are consistently exposed to drought, is their ability to form a sort of "memory". In a study, they found plants who had previously been exposed to drought were able to come up with a sort of strategy to minimize water loss and decrease water use. They found that plants who were exposed to drought conditions actually changed the way they regulated their stomata and what they called "hydraulic safety margin" so as to decrease the vulnerability of the plant. By changing the regulation of stomata and subsequently the transpiration, plants were able to function better in situations where the availability of water decreased.

Excessive water or water logging:

Waterlogging is the saturation of soil with water. Soil may be regarded as waterlogged when it is nearly saturated with water much of the time such that its air phase is restricted and anaerobic conditions prevail.

Waterlogging is overabundance water in the root zone joined by anaerobic conditions. The overabundance water represses vaporous trade with the air, and organic movement goes through accessible oxygen in the dirt air and water – likewise called anaerobiosis, anoxia or oxygen inadequacy.

- ❖ Soils don't need to be soaked (waterlogged) for gas trade to be repressed.
- ❖ These conditions influence rural plants severally:
- ❖ Supplement insufficiencies or poison levels
- ❖ Root passing
- ❖ Diminished development or demise of the plant.

Waterlogging that causes passing of more profound roots in winter can prompt droughting of plants in spring and early senescence of yearly harvests and fields in the Mediterranean environment of south-west of Western Australia.

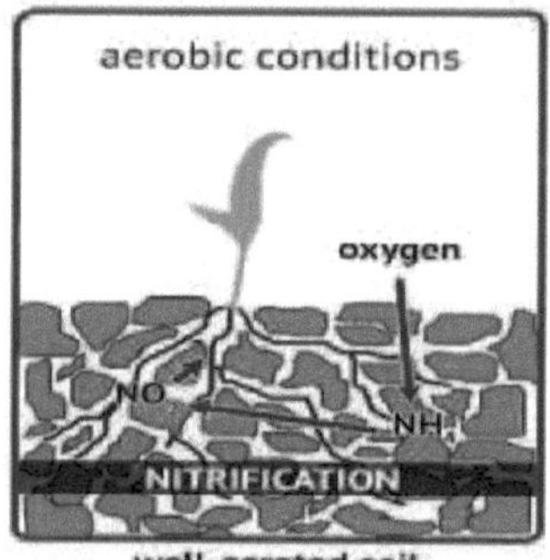

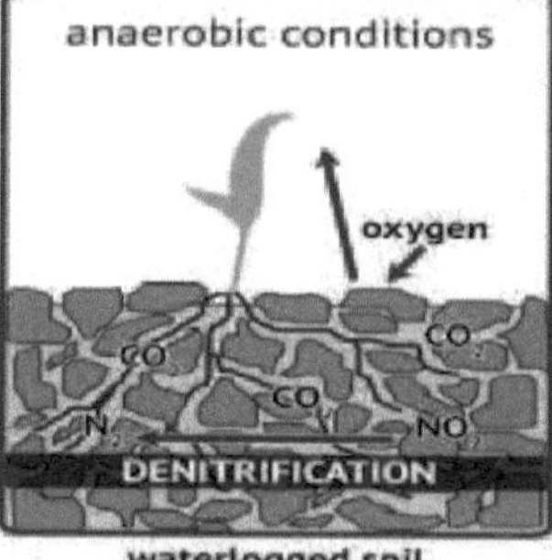

Figure 7: water logged soil causes the absence of oxygen

In extreme cases of prolonged waterlogging, anaerobiosis occurs, the roots of mesophytes suffer, and subsurface reducing atmosphere leads to such processes as denitrification, the reduction of iron and manganese oxides.

In agriculture, various crops need air (specifically, oxygen) to a greater or lesser depth in the soil. Waterlogging of the soil stops air getting in. Waterlogging brings down oxygen levels in the root zone, which lessens plant development. Waterlogging expands the decrease capability of the dirt and changes the synthetic harmony of numerous components which then, at that point enter the dirt water arrangement in their ionic structures.

Contingent upon soil type, this adjustment of synthetic balance can incorporate transient poison levels of some dirt supplements that are regularly protected when soil is uninhibitedly depleted. Instances of these are iron and manganese intensifies which can be changed over into free particles when decrease potential is high.

The decrease potential can require a few days or more to get back to ordinary levels after the waterlogged soil has depleted. Customary precipitation can cause continued waterlogging and seepage cycles, and the dirt can have high decrease potential for extensive stretches. Dirt that is over and again waterlogged for significant stretches regularly seems blue-dark, and may have a damp smell.

Waterlogging can prompt soil structure decay

Waterlogged soils will in general implode rapidly through scattering of mud particles. This is particularly the situation in soils with high sodium:calcium proportions; these dirts are regularly called dispersive or sodic soils. Waterlogged, non-dispersive soils can likewise lose soil structure by:

- ❖ Falling under their own load without a trace of extra strength created by the dirt dampness pull of unsaturated soil

- ❖ Aggravation by apparatus or domesticated animals when soaked at the surface.

Driving across waterlogged soils will harm soil structure. Limit the space of harm by utilizing a tramline or controlled traffic cultivating framework, which additionally further develops sellability during cultivating, treating and splashing procedure on waterlogged soils. The actual tramlines are defenseless to water disintegration, and we prescribe surface water the board constructions to limit the danger of harm.

Salinity

Soil salinity poses a global threat to world agriculture by reducing the yield of crops and ultimately the crop productivity in the salt affected areas. Salt stress reduces growth of crops and yield in many ways.

Most horticultural lands influenced by saltiness are situated in semi-parched or dry areas. Subsequently, the harm is strengthened by the synchronized activity of xerothermic perspectives, like aridity and high temperature (Huang et al., 2019). Salt pressure invigorates ROS age that harms biomolecules (e.g., lipids, proteins, and nucleic acids) and adjusts redox homeostasis (Kundu et al., 2018). Plants utilize various intends to detect, react, and adjust to changing the saline climate dependent on alterations in morpho-physiological characteristics and ionic, biochemical, and sub-atomic digestion systems which might additionally be improved by TU enlistment. Ongoing reports have proven the job of TU in working on the salt resistance and hidden components in many plants, including Indian mustard (Brassica juncea) (Srivastava et al., 2009), Halopyrum mucronatum (Khan and Ungar, 2001), wheat (Seleiman and Kheir, 2018), Aeluropus lagopodies (Gulzar and Khan, 2002), maize (Kaya et al., 2016), and mung bean

Figure 8: Salinity of soil

Two primary effects are imposed on crop plants by salt stress; **osmotic stress** and **ion toxicity**. The osmotic pressure under salinity stress in the soil solution exceeds the osmotic pressure in plant cells due to the presence of more salt, and thus, limits the ability of plants to take up water and minerals like K+ and Ca2+. These primary effects of salinity stress causes some secondary effects like assimilate production, reduced cell expansion and membrane function as well as decreased cytosolic metabolism.

<u>Toxicity</u>

The increased dependence of agriculture on chemical fertilizers and sewage waste water irrigation and rapid industrialization has added toxic metals to agriculture soils causing harmful effects on soil-plant environment system.

<u>Cold (low temperature)</u>

Cold stress as abiotic stress has proved to be the main abiotic stresses that decrease productivity of agricultural crops by affecting the quality of crops and their post-harvest life. Plants being immobile in nature are always busy to modify their mechanisms in order to prevent themselves from such stresses. In temperate conditions plants are encountered by chilling and freezing conditions that are very harmful to plants as stress. In order to adopt themselves, plants acquire chilling and freezing tolerance against such lethal cold stresses by a process called as acclimation.

Freezing stress

Figure 9: Cold stress or freezing stress in plants

However many important crops are still incompetent to the process of cold acclimation. The abiotic stress caused by cold affect the cellular functions of plants in every aspect. Several signal transduction pathways are there by which these cold stresses are transduced like components of ROS, protein kinase, protein phosphate, ABA and Ca2+, etc. and among these ABA proves to be best.

Heat (high temperature)

Increase in temperature throughout the globe has become a great concern, which not only affect the growth of plants but their productivity as well especially in agricultural crops plants. When plants encounter heat stress the percentage of seed germination, photosynthetic efficiency and yield declines.

Cool, chilling, and warmth stress during basic phases of plant advancement firmly affect formative falls of yield plants (Zhou et al., 2018). Temperature stress decreases the plant's ideal biochemical and physiological working by tweaking sub-atomic instruments (Djanaguiraman et al., 2018; Muhlemann et al., 2018; Takahashi and Shinozaki, 2019). For example, (Jatropha curcas) is a significant bioenergy crop, however it's capacity to deliver biofuel is frequently confined under cool pressure (Wang et al., 2013). Notwithstanding, TU (1.3 mM) seed preparing upgraded cold resistance by diminishing leaf senescence and keeping up with the general water content in jatropha plants developed at 4°C (Yadav et al., 2012).

Figure 10: Due to heat stress, plant petal cell's function lost

Under heat stress, during the reproductive growth period, the function of tapetal cells is lost, and the anther is dysplastic.

> **Water Deficit:**
>
> Is basically water reduction and cell dehydration in plants which causes stomata closure and eventually cell dies.

> **Salinity:**
>
> Is water reduction, cell dehydration and ion cytotoxicity which causes leaf expansion, cavitation and cell death.

> **Light stress:**
>
> Photo inhibition and ROS production causes reduced CO_2 fixation.

> **High temperature:**
>
> It effects membrane and protein destabilization which causes respiratory inhibition and cell death.

> **Chilling:**
>
> Effects membrane destabilization and membrane dysfunction.

> **Mineral nutrient deficiencies:**
>
> Reduced growth and ceases energy production.

<u>**Effects of abiotic stresses in plants:**</u>

Abiotic stress, as a natural part of every ecosystem, will affect organisms in a variety of ways. Although these effects may be either beneficial or detrimental, the location of the area is crucial in determining the extent of the impact that abiotic stress will have. The higher the latitude of the area affected, the greater the impact of abiotic stress will be on that area. So, a boreal forest is at the mercy of whatever abiotic stress factors may come along, while tropical zones are much less susceptible to such stressors.

Following effects have been observed in plants due to these abiotic stresses:

- There is the decreased photosynthesis and accumulation of reactive oxygen free radicals (ROS).

- There is the elevated transpiration and also the decrease in the carbon fixation process due to these abiotic stresses.

- Cell membrane of plants have been integrated due to abiotic stresses.

- There is the increase in hormonal imbalance and also the increased metal concentration in plants.

- Reduced nodulation and nitrogen fixation process in plants.

- Soil will contaminated with heavy metals.

- Physical properties such as soil texture, soil moisture content will be decreases due to Abiotic stresses.

<u>**Positive effects of Abiotic Stress in Plants**</u>

One example of a situation where abiotic stress plays a constructive role in an ecosystem is in natural wildfires. While they can be a human safety hazard, it is productive for these ecosystems to burn out every once in a while so that new organisms can begin to grow and thrive. Even though it is healthy for an ecosystem, a wildfire can still be considered an abiotic stressor, because it puts an obvious stress on individual organisms within the area. Every tree that is scorched and each bird nest that is devoured is a sign of the abiotic stress. On the larger scale, though, natural wildfires are positive manifestations of abiotic stress.

Lastly, abiotic stress has enabled species to grow, develop, and evolve, furthering natural selection as it picks out the weakest of a group of organisms. Both plants and animals have evolved mechanisms allowing them to survive extremes.

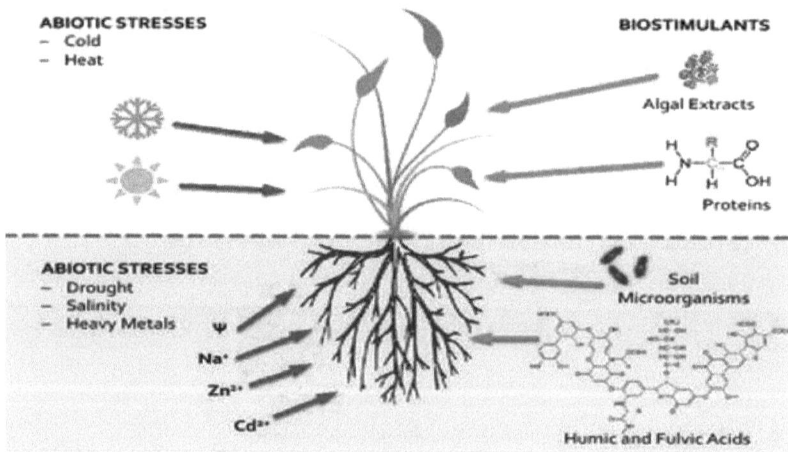

Figure 11: Abiotic stresses impacts in Plants.

What also needs to be taken into account when looking for benefits of abiotic stress, is that one phenomenon may not affect an entire ecosystem in the same way. While a flood will kill most plants living low on the ground in a certain area, if there is rice there, it will thrive in the wet conditions. Another example of this is in phytoplankton and zooplankton. The same types of conditions are usually considered stressful for these two types of organisms. They act very similarly when exposed to ultraviolet light and most toxins, but at elevated temperatures the phytoplankton reacts negatively, while the thermophilic zooplankton reacts positively to the increase in temperature.

The two may be living in the same environment, but an increase in temperature of the area would prove stressful only for one of the organisms. The most obvious detriment concerning abiotic stress involves farming. It has been claimed by one study that abiotic stress causes the most crop loss of any other factor and that most major crops are reduced in their yield by more than 50% from their potential yield. Because abiotic stress is widely considered a detrimental effect, the research on this branch of the issue is extensive.

Biotic Stress in Plants

Biotic stress is stress that occurs as a result of damage done to an organism by other living organisms, such as bacteria, viruses, fungi, parasites, beneficial and harmful insects, weeds, and cultivated or native plant.

Biotic components, or biotic elements, can be portrayed as any living part that influences another creature or shapes the ecosystem. This incorporates the two creatures that devour different life forms inside their environment, and the organic entity that is being burned-through. Biotic factors additionally incorporate human impact, microorganisms, and illness flare-ups. Each biotic factor needs a legitimate measure of energy and nourishment to work steadily.

Biotic Components

Biotic parts are regularly arranged into three primary classifications:

- ❖ Makers, also called autotrophs, convert energy (through the course of photosynthesis) into food.

- ❖ Shoppers, also called heterotrophs, rely on makers (and infrequently different customers) for food.

- ❖ Decomposers, also called detritivores, separate synthetics from makers and customers (typically anti-toxin) into easier structure which can be reused.

Difference from Abiotic Stress:

It is different from abiotic stress, which is the negative impact of non-living factors on the organisms such as temperature, sunlight, wind, salinity, flooding and drought. The types of biotic stresses imposed on an organism depend the climate where it lives as well as the species' ability to resist particular stresses. Biotic stress remains a broadly defined term and those who study it face many challenges, such as the greater difficulty in controlling biotic stresses in an experimental context compared to abiotic stress.

The damage caused by these various living and non-living agents can appear very similar. Even with close observation, accurate diagnosis can be difficult. For example, browning of leaves on an oak tree caused by drought stress may appear similar to leaf browning caused by oak wilt, a serious vascular disease caused by a fungus, or the browning caused by anthracnose, a fairly minor leaf disease.

All species are influenced by biotic factors in one way or another. For example, if the number of predators will increase, the whole food web will be affected as the population number of organisms that are lower in the food web will decrease due to predation. Similarly, when organisms have more food to eat, they will grow quicker and will be more likely to reproduce, so the population size will increase. Pathogens and disease outbreaks, however, are most likely to cause a decrease in population size. Humans make the most sudden changes in an environment (e.g. building cities and factories, disposing of waste into the water). These changes are most likely to cause a decrease in the population of any species due to the sudden appearance of pollutants.

Effect on Plant Growth:

Many biotic stresses affect photosynthesis, as chewing insects reduce leaf area and virus infections reduce the rate of photosynthesis per leaf area. Vascular-wilt fungi compromise the water transport and photosynthesis by inducing stomatal closure.

What are the plant defence mechanisms against biotic stress?

Plant defence mechanism is classified into two types:

1) Induced Structural Defence
2) Induced Chemical Defence

Induced Structural Defence

Induced structural defence involves:

- Cytoplasmic Defence
- Cell Wall Defence Structure
- Formation of Cork Layer

- Formation of Abscission Layer
- Deposition of Gums

Cytoplasmic Defence:

Some of the defence structures formed involve the cytoplasm of the cells under attack, and the process is called cytoplasmic defence reaction.

Cell Wall Defence Structure:

It involves morphological changes in the cell wall or change derived from the cell wall of the cell being invaded by the pathogen. Three main types of such structures have been observed in plant diseases.

> The outer layer of the cell wall of parenchyma cells coming in contact with incompatible bacteria swells and produced an amorphous, fibrillary material that surrounds and traps the bacteria and prevents them from multiplying.

> Cell walls thicken in response to several pathogens by producing what appears to be a cellulosic material.

> Callose papillae are deposited on the inner side of cell walls in response to invasion by fungal pathogen.

Formation of Cork Layers:

Infection by pathogens frequently induces plants to form several layers of Cork cells at the point of infection. The cork layers inhibit further invasion (entry) by the pathogen beyond the Initial lesion and also block the spread of any toxic substances that the Pathogen may secrete. Cork layers also stop the flow of Nutrients and water from the healthy to the infected area and deprive the pathogen of nourishment. The dead tissues, including the Pathogen, are thus delimited by the Cork layers.

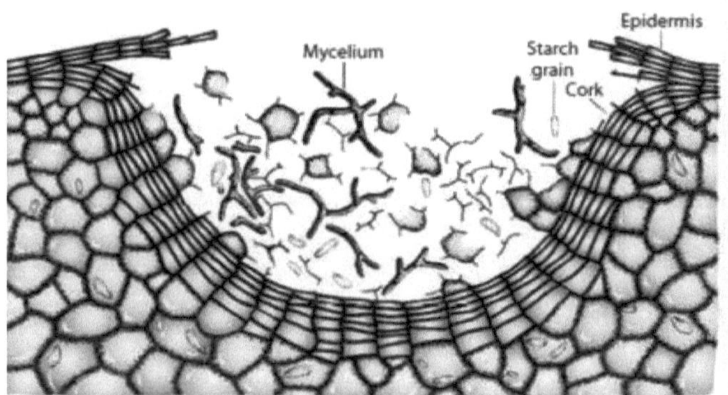

Figure 12: Formation of a cork layer on a potato tuber following infection with Rhizoctonia.

Formation of Abscission Layers:

It is a gap between host cell layers and devices for dropping –off older leaves and mature fruits. Plant may use this for defence mechanism to drop-off infected along with pathogen. An abscission layer consists of a gap formed between two circular layers of leaf cells surrounding the locus of infection. Upon infection, the middle lamella between these two layers of cells is dissolved throughout the thickness of the leaf, completely cutting off the central area of the infection from the rest of the leaf. This area shrivels, dies off, along with the pathogen. Thus, the plant, by discarding the infected area along with a few yet uninfected cells, protects the rest of the leaf tissue from being infected by the pathogen.

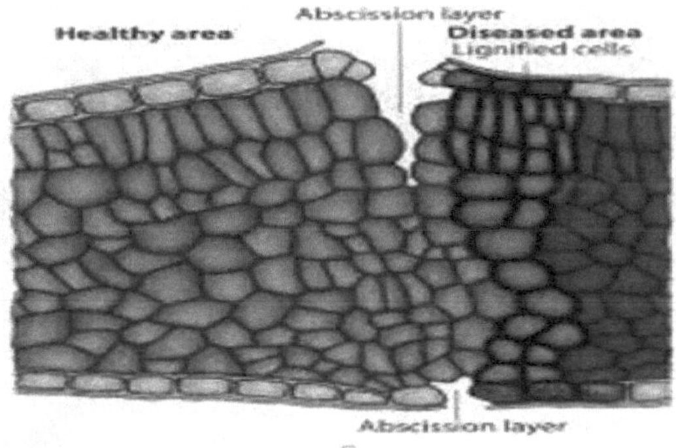

Figure 13: Formation of Abscission layer

Deposition of Gums:

Various types of gums are produced by many plants around lesions after infection by pathogens or injury. The defensive role of gums stems from the fact that they are deposited quickly in the intercellular spaces and within the cells surrounding the locus of infection, thus forming an impenetrable barrier that completely encloses the pathogen. The pathogen then becomes isolated, starves, and sooner or later dies.

Induced Chemical Defence

Induced Chemical Defence involves:

- Hyper Sensitive Response (HR)
- Pathogenesis-Related Proteins (PR-Proteins)
- Phytoalexins
- Systematic Acquired Resistance (SAR)

Hyper Sensitive Response (HR):

The hypersensitive response is localized death of host cells at site of infection. It is the result of a specific recognition of a pathogen attack by the host. The HR is considered to be a type of programmed cell death. The hypersensitive response is the plant defence response initiated by:

➢ The recognition by the plant of specific pathogen-produced signal molecules known as elicitors.

➢ Recognition of the elicitors by the host plant activates a cascade of biochemical reactions in the attacked and surrounding plant cells, leads to new or altered cell functions and to new or greatly activated defence – related compounds.

The most common new cell functions include:

- A rapid burst of reactive oxygen species, leading to a dramatic increase of oxidative reaction.

- Increased ion movement, especially of K+ and H+ through the cell membrane.

Pathogenesis-Related Proteins (PR-Proteins):

Pathogenesis related proteins, called PR- proteins- A group of plant coded proteins those are structurally diverse group toxic to invading pathogens. They are produced under stress. They are widely distributed in plants in trace amounts but are produced in high concentration following pathogen attack or stress.

Groups of PR-Proteins:

The better-known PR protein are:

- PR-1 proteins
- B-1,3- glucanases
- Chitinases
- Lysozyme
- PR-4 proteins
- Thaumatine like proteins

Phytoalexins:

This concept was given by **Borger** and **Muller**. **Paxton** (1981) defined phytoalexins are low molecular weight antimicrobial compounds which are synthesized by and accumulates in plant cells after microbial infection. Phytoalexins are toxic antimicrobial substances produced in appreciable amounts in plants only after stimulation by various types of phytopathogenic microorganisms or by chemical and mechanical injury. Most known phytoalexins are toxic to and inhibit the growth of fungi pathogenic to plants, but some are also toxic to bacteria, nematodes, and other organisms. Most phytoalexins elicitors are generally high molecular weight substances that are constituents of the fungal cell wall, such as glucans, chitosan, glycoproteins, and polysaccharides.

Systematic Acquired Resistance (SAR)

SAR is a mechanism of Induced defence that confers long lasting protection against a broad spectrum of microorganisms. It enhances resistance against subsequent array of pathogen. SAR takes 24-48hours to start and can last for months. It involves gene activation and a transmitted signal.

SAR refers to a distinct signal transduction process that plays an important role in the ability of plants to defend themselves against pathogens. After the formation of a necrotic lesion, either as a part of hypersensitive response (HR) or as a symptom of disease, the SAR process is activated. SAR activation results in the development of a broad- spectrum, systemic resistance.

Plants use design acknowledgment receptors to perceive moderated microbial marks. This acknowledgment triggers an insusceptible reaction. Plants likewise convey invulnerable receptors that perceive profoundly factor microbe effectors, these incorporate the NBS-LRR class of proteins. SAR is related with the enlistment of a wide scope of qualities (purported PR or "pathogenesis-related" qualities), and the enactment of SAR requires the amassing of endogenous salicylic corrosive (SA). The microorganism prompted SA signal enacts a sub-atomic sign transduction pathway that is distinguished by a quality called NIM1, NPR1 or SAI1 (three names for a similar quality) in the model hereditary framework Arabidopsis thaliana.

Surprisingly, the engineered fungicide acibenzolar-S-methyl isn't straightforwardly harmful to microorganisms, yet rather acts by prompting SAR in the yield plants to which it is applied. It is a propesticide — changed over in-vivo into 1, 2, 3-benzothiadiazole-7-carboxylic corrosive by methyl salicylate esterase. Field preliminaries have discovered that acibenzolar-S-methyl (otherwise called BSA)[citation needed] is compelling at controlling some plant illnesses, yet may have little impact on others, particularly contagious microorganisms which may not be entirely helpless to SAR.

Steps:

- Systemic acquired resistance is a local necrosis (HR) in the plant initiated by virus, fungus or bacteria triggers a local increase in salicylic acid (SA) accumulation and the formation of a phloem-mobile signal.
- The signal induces a SA and volatile methyl-SA accumulation in the systemic part of the plant which induce the synthesis of pathogenesis related proteins in the non-invaded parts of the plant.

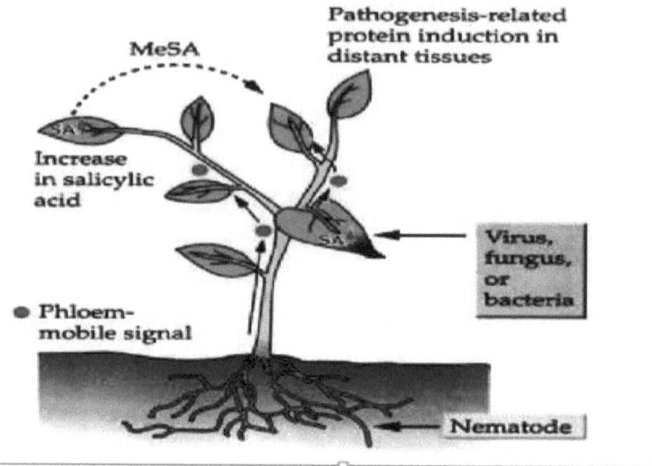

Figure 14: Systematic Acquired Resistance in plants.

How plants deal with stress???

Acclimation

"It is the adjustment of plant according to their environmental conditions".

When disease or another hard time hits, plants can react in a few different ways: adapt, avoid or die. They might develop resistance by becoming acclimated to growing and surviving despite the setback, or they might be susceptible and eventually die or fail to produce viable seeds. There are a wide range of ways that plants acclimate to stresses including:

> Changing their leaf size
> Developing antifreeze or heat-shock proteins,
> Adjusting the ions in their cells to compensate for dry soil.

Stress avoidance:

With very short growing seasons, they maximize moisture and temperature opportunities to mature, bloom and set seed quickly. Those seeds can lie dormant for many years, escaping potential threats until the right growing conditions happen again.

This is where the term hardy comes in, meaning the variety of the plant has an increased capacity for resilience despite environmental changes, such as cold winters.

Hormonal shifts in plants:

This is another way that they deal with stress—this is also seen in humans but in plants, different chemicals are emitted from different pores, depending on the stressor. Beneficial insects could be called in to attack a pest by these subtle yet powerful cues. Volatile organic compounds in the smell of freshly-cut grass is a familiar example of an aromatic cry for help.

How plants can avoid stress???

Eating healthy

Like humans, plants can benefit from "dietary changes". It is important that plants receive a healthy diet of essential macro nutrients and micronutrients, administered at key points of the plant's growth cycle.

For example, Phosphorus is critical to the development of seedlings, energy production and the establishment of a healthy and robust root system. Take the time to learn which nutrients are important at which growth stages and supplement the plant's diet with a sound fertility strategy.

Creating a stress-free environment:

There are several ways you can create a warm, safe and welcoming environment for your plants.

❖ Using a nutritional seed dressing can protect the seed against cracking and improve stand establishment while providing a timely boost of nutrients.

❖ Fortunately, there are now solutions that work at the plant's molecular level to prevent this temporary setback by inhibiting the production of ethylene in the plant – which reduces the impact of herbicide stress on the crop.

❖ Finally, taking appropriate steps to maintain soil health (like monitoring pH and mineral imbalances) should be part of an overall management practice.

Plants have co-evolved with their parasites for several hundred million years. This co-evolutionary process has resulted in the selection of a wide range of plant defences against microbial pathogens and herbivorous pests which act to minimise frequency and impact of attack. These defences include both physical and chemical adaptations, which may either be expressed constitutively, or in many cases, are activated only in response to attack. For example, utilization of high metal ion concentrations derived from the soil allow plants to reduce the harmful effects of biotic stressors (pathogens, herbivores etc.); meanwhile preventing the infliction of severe metal toxicity by way of safeguarding metal ion distribution throughout the plant with protective physiological pathways.

Such induced resistance provides a mechanism whereby the costs of defence are avoided until defence is beneficial to the plant. At the same time, successful pests and pathogens have evolved mechanisms to overcome both constitutive and induced resistance in their particular host species. In order to fully understand and manipulate plant biotic stress resistance, we require a detailed knowledge of these interactions at a wide range of scales, from the molecular to the community level.

✓ <u>**Inducible defence responses to insect herbivores**</u>

In order for a plant to defend itself against biotic stress, it must be able to differentiate between an abiotic and biotic stress. A plants response to herbivores starts with the recognition of certain chemicals that are abundant in the saliva of the herbivores. These compounds that trigger a response in plants are known as elicitors or herbivore-associated molecular patterns (HAMPs). These HAMPs trigger signalling pathways throughout the plant, initiating its defence mechanism and allowing the plant to minimise damage to other regions. These HAMPs trigger signalling pathways throughout the plant, initiating its defence mechanism and allowing the plant to minimise damage to other regions. Phloem feeders, like aphids, do not cause a great deal of mechanical damage to plants, but they are still regarded as pests and can seriously harm crop yields.

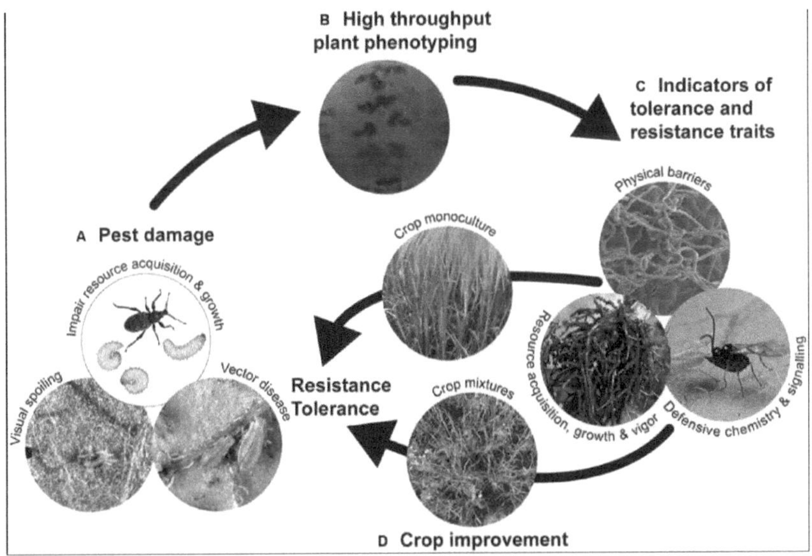

Figure 15: Inducible defense responses to insect herbivores.

Plants have developed a defence mechanism using salicylic acid pathway, which is also used in infection stress, when defending itself against phloem feeders. Plants perform a more direct attack on an insect's digestive system. The plants do this using proteinase inhibitors. These proteinase inhibitors prevent protein digestion and once in the digestive system of an insect, they bind tightly and specifically to the active site of protein hydrolysing enzymes such as trypsin and chymotrypsin. This mechanism is most likely to have evolved in plants when dealing with insect attack.

Plants detect elicitors in the insect's saliva. Once detected, a signal transduction network is activated. The presence of an elicitor causes an influx of Ca^{2+} ions to be released in to the cytosol. This increase in cytosolic concentration activates target proteins such as Calmodulin and other binding proteins. Downstream targets, such as phosphorylation and transcriptional activation of stimulus specific responses, are turned on by Ca^{2+} dependent protein kinases. In Arabidopsis, over expression of the IQD1 calmodulin-binding transcriptional regulator leads to inhibitor of herbivore activity. The role of calcium ions in this signal transduction network is therefore important.

Role of calcium ions

Calcium Ions also play a large role in activating a plants defensive response. When fatty acid amides are present in insect saliva, the mitogen-activated protein kinases (MAPKs) are activated. These genes when activated, play a role in the jasmonic acid pathway. The jasmonic acid pathway is also referred to as the Octadecanoid pathway. This pathway is vital for the activation of defence genes in plants. The production of jasmonic acid, a phytohormone, is a result of the pathway.

✓ Inducible defence responses to pathogens

Plants are equipped for recognizing intruders through the acknowledgment of non-self-signals regardless of the absence of a circulatory or resistant framework like those found in creatures. Frequently a plant's first line of protection against organisms happens at the plant cell surface and includes the identification of microorganism-related atomic examples (MAMPs). MAMPs incorporate nucleic acids normal to infections and endotoxins on bacterial cell layers which can be distinguished by particular example acknowledgment receptors. One more technique for identification includes the utilization of plant invulnerable receptors to distinguish effector particles delivered into plant cells by microorganisms. Location of these signs in tainted cells prompts an actuation of effector-set off resistance (ETI), a sort of intrinsic safe reaction.

Both the example acknowledgment insusceptibility (PTI) and effector-set off invulnerability (ETI) result from the upregulation of different safeguard components including cautious substance flagging mixtures. An increment in the creation of salicylic corrosive (SA) has been demonstrated to be prompted by pathogenic contamination. The increment in SA brings about the creation of pathogenesis related (PR) qualities which eventually increment plant protection from biotrophic and hemibiotrophic microbes. Expansions in jasmonic corrosive (JA) union close to the locales of microorganism contamination have additionally been depicted. This physiological reaction to build JA creation has been ensnared in the ubiquitination of jasmonate ZIM spaces (JAZ) proteins, which hinder JA flagging, prompting their corruption and a resulting expansion in JA enacted protection qualities.

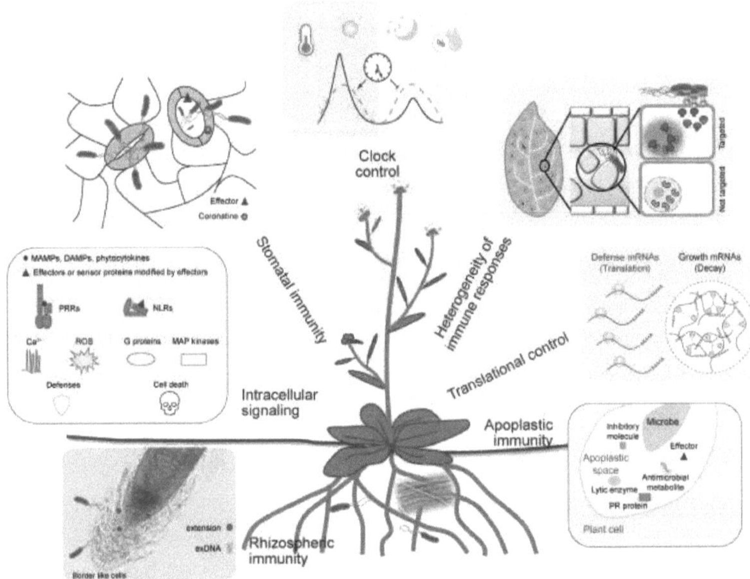

Figure 16: Inducible defense responses to pathogens

Studies in regards to the upregulation of guarded synthetics have affirmed the job of SA and JA in microbe protection. In investigations using Arabidopsis freaks with the bacterial NahG quality, which hinders the creation and gathering of SA, were demonstrated to be more powerless to microorganisms than the wild-type plants. This was thought to result from the powerlessness to create basic cautious systems including expanded PR quality articulation. Different investigations directed by infusing tobacco plants and Arabidopsis with salicylic corrosive brought about higher opposition of contamination by the hay and tobacco mosaic infections, showing a job for SA biosynthesis in decreasing viral replication. Also, concentrates on performed utilizing Arabidopsis with changed jasmonic corrosive biosynthesis pathways have demonstrated JA freaks to be at an expanded danger of contamination by soil microbes.

Alongside SA and JA, other cautious synthetic compounds have been embroiled in plant viral microbe protection including abscisic corrosive (ABA), gibberellic corrosive (GA), auxin, and peptide chemicals. The utilization of chemicals and natural invulnerability presents matches among creature and plant protections, however design set off resistance is thought to have emerged autonomously in each.

Applying new strategies

For agricultural crops, there is a growing category of plant enhancement technologies developed in the laboratory to help plants deal with stress – including cold temperatures, drought or wet conditions. Containing nutrients and bioactive compounds, these products stimulate desirable responses while down regulating undesirable responses at the plant's molecular level.

For example, these products may activate functions in the plant to increase water use efficiency; trigger earlier germination; promote enhanced root growth and much more. Taking advantage of this innovative technology early in the season will help protect plants against stress when they are at their most vulnerable.

Cross tolerance with abiotic stress

Proof shows that a plant going through numerous anxieties, both abiotic and biotic (generally microbe or herbivore assault), can create a positive result on plant execution, by decreasing their weakness to biotic pressure contrasted with how they react to individual burdens. The connection prompts a crosstalk between their separate chemical flagging pathways which will either incite or offend another rebuilding qualities hardware to expand resistance of safeguard reactions.

- ❖ Receptive oxygen species (ROS) are key flagging atoms delivered in light of biotic and abiotic stress cross resistance. ROS are created in light of biotic burdens during the oxidative burst.

- ❖ Double pressure forced by ozone (O3) and microbe influences resistance of yield and prompts modified host microorganism communication (Fuhrer, 2003). Modification in pathogenesis capability of bug because of O3 openness is of environmental and prudent importance.

Resistance to both biotic and abiotic stresses has been accomplished. In maize, rearing projects have prompted plants which are open minded to dry season and have extra protection from the parasitic weed *Striga hermonthica*.

<u>*Developmental and physiological mechanisms against environmental*</u>
<u>*stress*</u>

<u>Plants can modify their life cycles to avoid abiotic stress</u>

One way plants can adapt to extreme environmental conditions is through modification of their life cycles. For example, annual desert plants have short life cycles: they complete them during the periods when water is available, and are dormant (as seeds) during dry periods. Deciduous trees of the temperate zone shed their leaves before the winter so that sensitive leaf tissue is not damaged by cold temperatures. During less predictable stressful events (e.g., a summer of significant but erratic rainfall) the growth habits of some species may confer a degree of tolerance to these conditions. For example, plants that can grow and flower over an extended period (*indeterminate growth*) are often more tolerant to erratic environmental extremes than plants that develop preset numbers of leaves and flower over only very short periods (*determinate growth*).

<u>Phenotypic changes in leaf structure and behavior are important stress responses</u>

Because of their roles in photosynthesis, leaves (or their equivalent) are crucial to the survival of a plant. To function, leaves must be exposed to sunlight and air, but this also makes them particularly vulnerable to environmental extremes. Plants have thus evolved various mechanisms that enable them to avoid or mitigate the effects of abiotic extremes to leaves. Such mechanisms include changes in leaf area, leaf orientation, trachoma, and the cuticle.

Figure 17: Altered leaf shape can occur in response to environmental changes: leaf from outside (left) and inside (right) of a tree canopy

Relevance

The scope of abiotic and biotic factors spans across the entire biosphere, or global sum of all ecosystems. Such factors can have relevance for an individual within a species, its community or an entire population. For instance, disease is a biotic factor affecting the survival of an individual and its community. Temperature is an abiotic factor with the same relevance.

Some factors have greater relevance for an entire ecosystem. Abiotic and biotic factors combine to create a system or, more precisely, an ecosystem, meaning a community of living and non-living things considered as a unit. In this case, abiotic factors span as far as the pH of the soil and water, types of nutrients available and even the length of the day. Biotic factors such as the presence of autotrophs or self-nourishing organisms such as plants, and the diversity of consumers also affect an entire ecosystem.

Influencing Factors

Abiotic factors affect the ability of organisms to survive and reproduce. Abiotic limiting factors restrict the growth of populations. They help determine the types and numbers of organisms able to exist within an environment.

Figure 18: Abiotic stress as an influencing factors in Environment.

Biotic factors are living things that directly or indirectly affect organisms within an environment. This includes the organisms themselves, other organisms, interactions between living organisms and even their waste. Other biotic factors include parasitism, disease, and predation (the act of one animal eating another).

Interaction Examples

The significance of abiotic and biotic factors comes in their interaction with each other. For a community or an ecosystem to survive, the correct interactions need to be in place.

A simple example would be of abiotic interaction in plants. Water, sunlight and carbon dioxide are necessary for plants to grow. The biotic interaction is that plants use water, sunlight and carbon dioxide to create their own nourishment through a process called photosynthesis.

On a larger scale, abiotic interactions refer to patterns such as climate and seasonality. Factors such as temperature, humidity and the presence or absence of seasons affect the ecosystem. For instance, some ecosystems experience cold winters with a lot of snow. An animal such as a fox within this ecosystem adapts to these abiotic factors by growing a thick, <u>white</u>-coloured coat in the winter.

Decomposers such as bacteria and fungi are examples of biotic interactions on such a scale. Decomposers function by breaking down dead organisms. This process returns the basic components of the organisms to the soil, allowing them to be reused within that ecosystem.

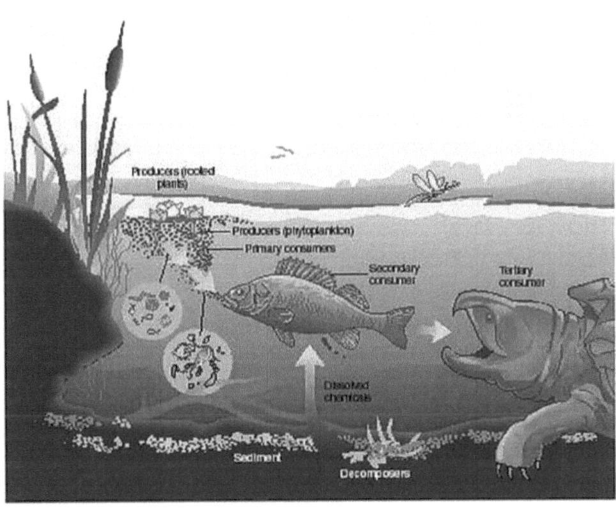

- Producers, i.e. autotrophs: e.g. plants, convert the energy [from photosynthesis (the transfer of sunlight, water, and carbon dioxide into energy), or other sources such as hydrothermal vents] into food.

- Consumers, i.e. heterotrophs: e.g. animals, depend upon producers (occasionally other consumers) for food.

- Decomposers, i.e. detritivores: e.g. fungi and bacteria, break down chemicals from producers and consumers (usually not living) into simpler form which can be reused.

Conclusion:

It is expected that world's temperature will increment by 3–5°C in the coming 50–100 years. As there is consistent expansion in temperature and lopsided precipitation the progressions of flood and dry season is consistently in thought. The anthropogenic exercises like inordinate composts, improper water system and double-dealing of metal assets can prompt salt pressure generally Biotic and Abiotic Stresses in Plants. A dangerous atmospheric devation prompts the simultaneousness of various abiotic and biotic anxieties, along these lines influencing horticultural usefulness. Event of abiotic stresses can change plant–bother cooperations by upgrading host plant vulnerability to pathogenic organic entities, creepy crawlies, and by diminishing serious capacity with weeds. Despite what might be expected, a few bugs might change plant reaction to abiotic stress factors. Along these lines, orderly investigations are vital to comprehend the impact of simultaneous abiotic and biotic pressure conditions on crop efficiency.

It is the obligation of plant reproducers to foster pressure lenient cultivars to get food security and to guarantee wellbeing to the ranchers. Atomic work is to be done at the hereditary level to foster components in plants to keep them from various kinds of stress conditions. Except if responsive systems are not created against biotic and abiotic stresses, the plants will persistently exposed to such anxieties and at last will demonstrate an extraordinary danger to world farming.

<u>References</u>

1. "Abiotic Stress". Biology Online. Archived from the original on 13 June 2008. Retrieved 2008-05-04.

2. Jump up to:[a] [b] Vinebrooke, Rolf D.; et al. (2004). "Impacts of multiple stressors on biodiversity and ecosystem functioning: the role of species co-tolerance". OIKOS. **104** (3): 451–457. doi:10.1111/j.0030-1299.2004.13255.x.

3. Gao, Ji-Ping; et al. (2007). "Understanding Abiotic Stress Tolerance Mechanisms: Recent Studies on Stress Response in Rice". Journal of Integrative Plant Biology. **49** (6): 742–750. doi:10.1111/j.1744-7909.2007.00495.x.

4. Jump up to:[a] [b] Mittler, Ron (2006). "Abiotic stress, the field environment and stress combination". Trends in Plant Science. **11** (1): 15–19. doi:10.1016/j.tplants.2005.11.002. PMID 16359910.

5. Jump up to:[a] [b] [c] [d] [e] [f] [g] [h] [i] Palta, Jiwan P. and Farag, Karim. "Methohasds for enhancing plant health, protecting plants from biotic and abiotic stress related injuries and enhancing the recovery of plants injured as a result of such stresses." United States Patent 7101828, September 2006.

6. Voesenek, LA; Bailey-Serres, J (April 2015). "Flood adaptive traits and processes: an overview". The New Phytologist. **206** (1): 57–73. doi:10.1111/nph.13209. PMID 25580769.

7. Sasidharan, R; Hartman, S; Liu, Z; Martopawiro, S; Sajeev, N; van Veen, H; Yeung, E; Voesenek, LACJ (February 2018). "Signal Dynamics and Interactions during Flooding Stress". Plant Physiology. **176** (2): 1106–1117. doi:10.1104/pp.17.01232. PMC 5813540. PMID 29097391.

8. Jump up to:[a] [b] Wolfe, A. "Patterns of biodiversity." Ohio State University, 2007.

9. Jump up to:[a] [b] [c] Brussaard, Lijbert; de Ruiter, Peter C.; Brown, George G. (2007). "Soil biodiversity for agricultural sustainability". Agriculture, Ecosystems and Environment. **121** (3): 233–244. doi:10.1016/j.agee.2006.12.013.

10. Jump up to:[a] [b] Roelofs, D.; et al. (2008). "Functional ecological genomics to demonstrate general and specific responses to abiotic stress". Functional Ecology. **22**: 8–18. doi:10.1111/j.1365-2435.2007.01312.x.

11. Wang, W.; Vinocur, B.; Altman, A. (2007). "Plant responses to drought, salinity and extreme temperatures towards genetic engineering for stress tolerance". Planta. **218** (1): 1–14. doi:10.1007/s00425-003-1105-5. PMID 14513379. S2CID 24400025.

12. Cramer, Grant R; Urano, Kaoru; Delrot, Serge; Pezzotti, Mario; Shinozaki, Kazuo (2011-11-17). "Effects of abiotic stress on plants: a systems biology perspective". BMC Plant Biology. **11**: 163. doi:10.1186/1471-2229-11-163. ISSN 1471-2229. PMC 3252258. PMID 22094046.

13. Conde, Artur (2011). "Membrane Transport, Sensing and Signaling in Plant Adaptation to Environmental Stress" (PDF). Plant & Cell Physiology. **52** (9): 1583–1602. doi:10.1093/pcp/pcr107. PMID 21828102 – via Google Scholar.

14. Maestre, Fernando T.; Cortina, Jordi; Bautista, Susana (2007). "Mechanisms underlying the interaction between Pinus halepensis and the native late-successional shrub Pistacia lentiscus in a semi-arid plantation". Ecography. **27** (6): 776–786. doi:10.1111/j.0906-7590.2004.03990.x.

15. Palm, Brady; Van Volkenburgh (2012). "Serpentine tolerance in Mimuslus guttatus does not rely on exclusion of magnesium". Functional Plant Biology. **39** (8): 679–688. doi:10.1071/FP12059. PMID 32480819.

16. Singh, Samiksha; Parihar, Parul; Singh, Rachana; Singh, Vijay P.; Prasad, Sheo M. (2016). "Heavy Metal Tolerance in Plants: Role of Transcriptomics, Proteomics, Metabolomics, and Ionomics". Frontiers in Plant Science. **6**: 1143. doi:10.3389/fpls.2015.01143. ISSN 1664-462X. PMC 4744854. PMID 26904030.

17. Savvides, Andreas (December 15, 2015). "Chemical Priming of Plants Against Multiple Abiotic Stresses: Mission Possible?". Trends in Plant Science. **21** (4): 329–340. doi:10.1016/j.tplants.2015.11.003. hdl:10754/596020. PMID 26704665. Retrieved March 10, 2016.

18. Gull, Audil; Lone, Ajaz Ahmad; Wani, Noor Ul Islam (2019-10-07). "Biotic and Abiotic Stresses in Plants". Abiotic and Biotic Stress in Plants. doi:10.5772/intechopen.85832. ISBN 978-1-78923-811-2.

19. Jump up to:[a][b][c][d][e] Yadav, Summy; Modi, Payal; Dave, Akanksha; Vijapura, Akdasbanu; Patel, Disha; Patel, Mohini (2020-06-17). "Effect of Abiotic Stress on Crops". Sustainable Crop Production. doi:10.5772/intechopen.88434. ISBN 978-1-78985-317-9.

20. Martinez-Beltran J, Manzur CL. (2005). Overview of salinity problems in the world and FAO strategies to address the problem. Proceedings of the international salinity forum, Riverside, California, April 2005, 311–313.

21. Jump up to:[a] [b] [c] Neto, Azevedo; De, André Dias; Prisco, José Tarquinio; Enéas-Filho, Joaquim; Lacerda, Claudivan Feitosa de; Silva, José Vieira; Costa, Paulo Henrique Alves da; Gomes-Filho, Enéas (2004-04-01). "Effects of salt stress on plant growth, stomatal response and solute accumulation of different maize genotypes". Brazilian Journal of Plant Physiology. **16** (1): 31–38. doi:10.1590/S1677-04202004000100005. ISSN 1677-0420.

22. Zhu, J.-K. (2001). Plant Salt Stress. eLS.

23. Lu. Congming, A. Vonshak. (2002). Effects of salinity stress on photosystem II function in cyanobacterial Spirulina platensis cells. Physiol. Plant 114 405-413.

24. Lei, Gang; Shen, Ming; Li, Zhi-Gang; Zhang, Bo; Duan, Kai-Xuan; Wang, Ning; Cao, Yang-Rong; Zhang, Wan-Ke; Ma, Biao (2011-10-01). "EIN2 regulates salt stress response and interacts with a MA3 domain-containing protein ECIP1 in Arabidopsis". Plant, Cell & Environment. **34** (10): 1678–1692. doi:10.1111/j.1365-3040.2011.02363.x. ISSN 1365-3040. PMID 21631530.

25. Raghothama, K. G. (1999-01-01). "Phosphate Acquisition". Annual Review of Plant Physiology and Plant Molecular Biology. **50** (1): 665–693. doi:10.1146/annurev.arplant.50.1.665. PMID 15012223.

26. Rubio, Vicente; Linhares, Francisco; Solano, Roberto; Martín, Ana C.; Iglesias, Joaquín; Leyva, Antonio; Paz-Ares, Javier (2001-08-15). "A conserved MYB transcription factor involved in phosphate starvation signaling both in vascular plants and in unicellular algae". Genes & Development. **15** (16): 2122–2133. doi:10.1101/gad.204401. ISSN 0890-9369. PMC 312755. PMID 11511543.

27. Jump up to:[a] [b] Pant, Bikram Datt; Burgos, Asdrubal; Pant, Pooja; Cuadros-Inostroza, Alvaro; Willmitzer, Lothar; Scheible, Wolf-Rüdiger (2015-04-01). "The transcription factor PHR1 regulates lipid remodeling and triacylglycerol accumulation in Arabidopsis thaliana during phosphorus starvation". Journal of Experimental Botany. **66** (7): 1907–1918. doi:10.1093/jxb/eru535. ISSN 0022-0957. PMC 4378627. PMID 25680792.

28. Pant, Bikram-Datt; Pant, Pooja; Erban, Alexander; Huhman, David; Kopka, Joachim; Scheible, Wolf-Rüdiger (2015-01-01). "Identification of primary and secondary metabolites with phosphorus status-dependent abundance in Arabidopsis, and of the transcription factor PHR1 as a major regulator of metabolic changes during phosphorus

limitation". Plant, Cell & Environment. **38** (1): 172–187. doi:10.1111/pce.12378. ISSN 1365-3040. PMID 24894834.

29. Ogbaga, Chukwuma C.; Athar, Habib-ur-Rehman; Amir, Misbah; Bano, Hussan; Chater, Caspar C.C.; Jellason, Nugun P. (July 2020). "Clarity on frequently asked questions about drought measurements in plant physiology". Scientific African. **8**: e00405. doi:10.1016/j.sciaf.2020.e00405.

30. Jump up to:[a] [b] González-Villagra, Jorge; Rodrigues-Salvador, Acácio; Nunes-Nesi, Adriano; Cohen, Jerry D.; Reyes-Díaz, Marjorie M. (March 2018). "Age-related mechanism and its relationship with secondary metabolism and abscisic acid in Aristotelia chilensis plants subjected to drought stress". Plant Physiology and Biochemistry. **124**: 136–145. doi:10.1016/j.plaphy.2018.01.010. ISSN 0981-9428. PMID 29360623.

31. Jump up to:[a] [b] [c] Tombesi, Sergio; Frioni, Tommaso; Poni, Stefano; Palliotti, Alberto (June 2018). "Effect of water stress "memory" on plant behavior during subsequent drought stress". Environmental and Experimental Botany. **150**: 106–114. doi:10.1016/j.envexpbot.2018.03.009. ISSN 0098-8472.

32. Zargar, Sajad Majeed; Nagar, Preeti; Deshmukh, Rupesh; Nazir, Muslima; Wani, Aijaz Ahmad; Masoodi, Khalid Zaffar; Agrawal, Ganesh Kumar; Rakwal, Randeep (October 2017). "Aquaporins as potential drought tolerance inducing proteins: Towards instigating stress tolerance". Journal of Proteomics. **169**: 233–238. doi:10.1016/j.jprot.2017.04.010. ISSN 1874-3919. PMID 28412527.

33. Goncalves-Alvim, Silmary J.; Fernandez, G. Wilson (2001). "Biodiversity of galling insects: historical, community and habitat effects in four neotropical savannas". Biodiversity and Conservation. **10**: 79–98. doi:10.1023/a:1016602213305. S2CID 37515138.

34. Peterson, R. K. D.; Higley, L. G., eds. (2001). Biotic Stress and Yield Loss. CRC Press. ISBN 978-0849311451.

35. Pinter, P. J.; Hatfield, J. L.; Schepers, J. S.; Barnes, E. M.; Moran, M. S.; Daughtry, C. S.T.; Upchurch, D. R. (2003). "Remote Sensing for Crop Management". Photogrammetric Engineering & Remote Sensing. **69** (6): 647–664. doi:10.14358/PERS.69.6.647.

36. Poschenrieder, C.; Tolrà, R.; Barceló, J. (2006). "Can metals defend plants against biotic stress?". Trends in Plant Science. **11** (6): 288–295. doi:10.1016/j.tplants.2006.04.007. PMID 16697693.

37. Raju, N. J.; Gossel, W.; Ramanathan, A.; Sudhakar, M., eds. (2015). Management of Water, Energy and Bio-resources in the Era of Climate Change: Emerging Issues and Challenges. doi:10.1007/978-3-319-05969-3. ISBN 978-3319059686. S2CID 132165592.

38. Rejeb, I. B.; Pastor, V.; Mauch-Mani, B. (2014). "Plant Responses to Simultaneous Biotic and Abiotic Stress: Molecular Mechanisms". Plants. **3** (4): 458–475. doi:10.3390/plants3040458. PMC 4844285. PMID 27135514.

39. Roberts, M. (2013). "Preface: Induced Resistance to biotic stress". Journal of Experimental Botany. **64** (5): 1235–1236. doi:10.1093/jxb/ert076. PMID 23616991.

40. Yadav, B. K. V. (2012). "Abiotic stress". Forestrynepal.org. Retrieved 3 December 2015.

41. Flynn, P. (2003). "Biotic vs. Abiotic - Distinguishing Disease Problems from Environmental Stresses". ISU Entomology. Retrieved 16 May 2013.

42. Fuller, V. L.; Lilley, C. J.; Urwin, P. E. (2008). "Nematode resistance". New Phytologist. **180** (1): 27–44. doi:10.1111/j.1469-8137.2008.02508.x. PMID 18564304.

43. Garrett, K. A.; Dendy, S. P.; Frank, E. E.; Rouse, M. N.; Travers, S. E. (2006). "Climate Change Effects on Plant Disease: Genomes to Ecosystems" (PDF). Annual Review of Phytopathology. **44**: 489–509. doi:10.1146/annurev.phyto.44.070505.143420. hdl:2097/2379. PMID 16722808.

44. Karim, S. (2007). Exploring plant tolerance to biotic and abiotic stresses (PDF) (Thesis). ISBN 978-9157673572.

45. Perez, I. B.; Brown, P. J. (2014). "The role of ROS signaling in cross-tolerance: from model to crop". Front. Plant Sci. **5**: 754. doi:10.3389/fpls.2014.00754. PMC 4274871. PMID 25566313

46. Atkinson, N. J.; Urwin, P. E. (2012). "The interaction of plant biotic and abiotic stresses: from genes to the field". Journal of Experimental Botany. **63** (10): 3523–3543. doi:10.1093/jxb/ers100. PMID 22467407.

47. Carris, L. M.; Little, C. R.; Stiles, C. M. (2012). "Introduction to Fungi". www.apsnet.org. doi:10.1094/PHI-I-2012-0426-01 (inactive 31 May 2021). Archived from the original on 2015-11-10. Retrieved 11 March 2016.

48. Flexas, J.; Loreto, F.; Medrano, H., eds. (2012). Terrestrial Photosynthesis In A Changing Environment: A Molecular, Physiological, and Ecological Approach. CUP. ISBN 978-0521899413.

49. Suda, Kenichi; Katagiri, Fumiaki (2010). "Comparing signaling mechanisms engaged in pattern-triggered and effector-triggered immunity". Current Opinion in Plant Biology. 13(4): 459–465. doi:10.1016/j.pbi.2010.04.006. PMID 20471306.

50. ^ Jump up to:[a] [b] [c] Bari, Rajendra; Jones, Jonathan D. G. (2009). "Role of plant hormones in plant defence responses". Plant Molecular Biology. 69 (4): 473–488. doi:10.1007/s11103-008-9435-0. PMID 19083153. S2CID 28385498.

51. ^ Jump up to:[a] [b] [c] Halim, V. A.; Vess, A.; Scheel, D.; Rosahl, S. (2006). "The Role of Salicylic Acid and Jasmonic Acid in Pathogen Defence". Plant Biology. 8 (3): 307–313. doi:10.1055/s-2006-924025. PMID 16807822. S2CID 28317435.

52. ^ Jump up to:[a] [b] Vlot, A. Corina; Dempsey, D'Maris Amick; Klessig, Daniel F. (2009). "Salicylic Acid, a Multifaceted Hormone to Combat Disease". Annual Review of Phytopathology. 47: 177–206. doi:10.1146/annurev.phyto.050908.135202. PMID 19400653.

53. ^ Van Huijsduijnen, R. A. M. H.; Alblas, S. W.; De Rijk, R. H.; Bol, J. F. (1986). "Induction by Salicylic Acid of Pathogenesis-related Proteins and Resistance to Alfalfa Mosaic Virus Infection in Various Plant Species". Journal of General Virology. 67 (10): 2135–2143. doi:10.1099/0022-1317-67-10-2135

54. Taiz Lincoln, Zeiger Eduardo, Møller Ian Max, Murphy Angus (2015). Plant Physiology and Development. USA: Sinauer Associations, Inc. p. 706. ISBN 9781605352558.

55. ^ Jump up to:[a] [b] Spoel, Steven H.; Dong, Xinnian (2012). "How do plants achieve immunity? Defence without specialized immune cells". Nature Reviews Immunology. 12 (2): 89–100. doi:10.1038/nri3141. PMID 22273771. S2CID 205491561.

56. Boller, T; He, SY (2009). "Innate immunity in plants: an arms race between pattern recognition receptors in plants and effectors in microbial pathogens". Science. 324 (5928): 742–4. Bibcode:2009Sci...324.742B. Doi: 10.1126/science.1171647. PMC 2729760. PMID 19423812

57. Balachandran; et al. (1997). "Concepts of plant biotic stress. Some insights into the stress physiology of virus infected plants, from the perspective of photosynthesis". Physiologia Plantarum. 100 (2): 203–213. doi:10.1111/j.1399-3054.1997.tb04776.

58. Ausubel FM (October 2005). "Are innate immune signaling pathways in plants and animals conserved?". Nature Immunology. **6** (10): 973–9. doi:10.1038/ni1253. PMID 16177805. S2CID 7451505.

59. Jump up to:[a] [b] [c] [d] Ryals, J., U. Neuenschwander, M. Willits, A. Molina, H. Steiner, and M. Hunt. 1996. Systemic Acquired Resistance. Plant Cell 8:1809–1819.

60. Song WY, Wang GL, Chen LL, Kim HS, Pi LY, Holsten T, Gardner J, Wang B, Zhai WX, Zhu LH, Fauquet C, Ronald P (December 1995). "A receptor kinase-like protein encoded by the rice disease resistance gene, Xa21". Science. **270** (5243): 1804–6. doi:10.1126/science.270.5243.1804. PMID 8525370. S2CID 10548988.

61. ^ Gómez-Gómez L, Boller T (June 2000). "FLS2: an LRR receptor-like kinase involved in the perception of the bacterial elicitor flagellin in Arabidopsis". Molecular Cell. **5** (6): 1003–11. doi:10.1016/S1097-2765(00)80265-8. PMID 10911994.

62. Jeschke, Peter (2016). "Propesticides and their use as agrochemicals". Pest Management Science. **72** (2): 210–225. doi:10.1002/ps.4170. PMID 26449612.

63. Vallad, Gary E.; Goodman, Robert M. (2004). "Systemic Acquired Resistance and Induced Systemic Resistance in Conventional Agriculture". Crop Science. **44** (6): 1920–1934. doi:10.2135/cropsci2004.1920. ISSN 1435-0653. Retrieved 2020-11-27

64. Dai, A. (2011). Drought under global warming: a review. Wiley Interdiscip. Rev. Clim. Chang. 2, 45–65. doi: 10.1002/wcc.81

65. Dangol, S., Chen, Y., Hwang, B. K., Jwa, N.-S. (2019). Iron-and reactive oxygen species-dependent ferroptotic cell death in rice-Magnaporthe oryzae interactions. Plant Cell 31, 189–209. doi: 10.1105/tpc.18.00535

66. Demidchik, V., Cuin, T. A., Svistunenko, D., Smith, S. J., Miller, A. J., Shabala, S., et al. (2010). Arabidopsis root K+-efflux conductance activated by hydroxyl radicals: single-channel properties, genetic basis and involvement in stress-induced cell death. J Cell Sci 123, 1468–1479. doi: 10.1242/jcs.064352

67. Dias, M. C., Correia, S., Serôdio, J., Silva, A. M. S., Freitas, H., Santos, C. (2018). Chlorophyll fluorescence and oxidative stress endpoints to discriminate olive cultivars tolerance to drought and heat episodes. Sci. Hortic. (Amsterdam). 231, 31–35. doi: 10.1016/j.scienta.2017.12.007

68. Djanaguiraman, M., Perumal, R., Jagadish, S. V. K., Ciampitti, I. A., Welti, R., Prasad, P. V. V., et al. (2018). Sensitivity of sorghum pollen and pistil to high-temperature stress. Plant. Cell Environ. 41, 1065–1082. doi: 10.1111/pce.13089.

69. Couto, R., Comin, J. J., Souza, M., Ricachenevsky, F. K., Lana, M., Gatiboni, L., et al. (2018). Should heavy metals be monitored in foods derived from soils fertilized with animal waste? *Front. Plant Sci.* 9, 732. doi: 10.3389/fpls.2018.00732

70. Czarnocka, W., Karpiński, S. (2018). Friend or foe? Reactive oxygen species production, scavenging and signaling in plant response to environmental stresses. *Free Radic. Biol. Med.* 122, 4–20. doi: 10.1016/j.freeradbiomed.2018.01.011.

71. Dos Reis, S. P., Lima, A. M., de Souza, C. R. B. (2012). Recent molecular advances on downstream plant responses to abiotic stress. *Int. J. Mol. Sci.* 13, 8628–8647. doi: 10.3390/ijms13078628

72. Ceusters, N., Van den Ende, W., Ceusters, J. (2016). "Exploration of sweet immunity to enhance abiotic stress tolerance in plants," In *Lessons from CAM," in Progress in Botany*, vol. 78. (Switzerland: Springer), 145–166. doi: 10.1007/124_2016_1

73. Borisova, G., Chukina, N., Maleva, M., Kumar, A., Prasad, M. N. V. (2016). Thiols as biomarkers of heavy metal tolerance in the aquatic macrophytes of Middle Urals, Russia. *Int. J. Phytoremediation* 18, 1037–1045. doi: 10.1080/15226514.2016.1183572

74. Babar, B. H. (2013). *Improving drought tolerance in maize (Zea mays L.) by exogenous application of thiourea.* (Doctoral dissertation, University of Agriculture, Faisalabad,

Pakistan)   Available
at http://prr.hec.gov.pk/jspui/bitstream/123456789/1098/1/1971S.pdf .

75. Bartlett, M. K., Detto, M., Pacala, S. W. (2019). Predicting shifts in the functional composition of tropical forests under increased drought and CO 2 from trade-offs among plant hydraulic traits. *Ecol. Lett.* 22, 67–77. doi: 10.1111/ele.13168

76. Bartwal, A., Mall, R., Lohani, P., Guru, S. K., Arora, S. (2013). Role of secondary metabolites and brassinosteroids in plant defense against environmental stresses. *J. Plant Growth Regul.* 32, 216–232. doi: 10.1007/s00344-012-9272-x

77. Beuchat, J., Scacchi, E., Tarkowska, D., Ragni, L., Strnad, M., Hardtke, C. S. (2010). BRX promotes Arabidopsis shoot growth. *New Phytol.* 188, 23–29. doi: 10.1111/j.1469-8137.2010.03387.x

78. Huang, M., Chai, L., Jiang, D., Zhang, M., Zhao, Y., Huang, Y. (2019). Increasing aridity affects soil archaeal communities by mediating soil niches in semi-arid regions. *Sci. Total Environ.* 647, 699–707. doi: 10.1016/j.scitotenv.2018.07.305

79. Isayenkov, S., Maathuis, F. J. M. (2019). Plant salinity stress; many unanswered questions remain. *Front. Plant Sci.* 10, 1–11. doi: 10.3389/fpls.2019.00080

80. Khan, M. A., Ungar, I. A. (2001). Alleviation of salinity stress and the response to temperature in two seed morphs of Halopyrum mucronatum (Poaceae). *Aust. J. Bot.* 49, 777–783. doi: 10.1071/BT01014.

81. Kundu, P., Gill, R., Ahlawat, S., Anjum, N. A., Sharma, K. K., Ansari, A. A., et al. (2018). "Targeting the redox regulatory mechanisms for abiotic stress tolerance in crops," in *Biochemical, Physiological and Molecular Avenues for Combating Abiotic Stress Tolerance in Plants* (Elsevier Academic Press), 151–220.

82. Mantri, N., Patade, V., Penna, S., Ford, R., Pang, E. (2012). "Abiotic stress responses in plants: present and future," in *Abiotic stress responses in plants* (New York: Springer), 1–19.

83. Narendrula-Kotha, R., Theriault, G., Mehes-Smith, M., Kalubi, K., Nkongolo, K. (2019). Metal toxicity and resistance in plants and microorganisms in terrestrial ecosystems. In *Reviews of Environmental Contamination and Toxicology* (Switzerland: Springer), 249, 1–27.

84. Pasala, P. S. M. R. K. (2017). *Abiotic Stress Management for Resilient Agriculture.* (Singapore: Springer), 233–260.

85. Patade, V. Y., Khatri, D., Manoj, K., Kumari, M., Ahmed, Z. (2012). Cold tolerance in thiourea primed capsicum seedlings is associated with transcript regulation of stress responsive genes. *Mol. Biol. Rep.* 39, 10603–10613.

86. Proshad, R., Kormoker, T., Mursheed, N., Islam, M. M., Bhuyan, M. I., Islam, M. S., Mithu, T. N. (2018). Heavy metal toxicity in agricultural soil due to rapid industrialization in Bangladesh: a review. *Int. J. Adv. Geosci.* 6 (1), 83–88.

87. Rehman, K., Fatima, F., Waheed, I., Akash, M. S. H. (2018). Prevalence of exposure of heavy metals and their impact on health consequences. *J. Cell. Biochem.* 119 (1), 157–184.

88. Seleiman, M. F., Kheir, A. M. S. (2018). Saline soil properties, quality and productivity of wheat grown with bagasse ash and thiourea in different climatic zones. *Chemosphere* 193, 538–546.

Thank You

I **want** morebooks!

Buy your books fast and straightforward online - at one of world's fastest growing online book stores! Environmentally sound due to Print-on-Demand technologies.

Buy your books online at
www.morebooks.shop

Kaufen Sie Ihre Bücher schnell und unkompliziert online – auf einer der am schnellsten wachsenden Buchhandelsplattformen weltweit! Dank Print-On-Demand umwelt- und ressourcenschonend produziert.

Bücher schneller online kaufen
www.morebooks.shop

KS OmniScriptum Publishing
Brivibas gatve 197
LV-1039 Riga, Latvia
Telefax: +371 686 204 55

info@omniscriptum.com
www.omniscriptum.com

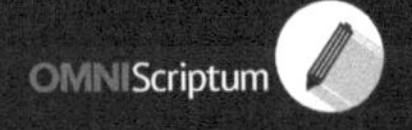

Printed by Books on Demand GmbH, Norderstedt / Germany